Vintage Snowmobilia

A Guide to Snowmobile Collectibles

Jon D. Bertolino

Iconografix

Iconografix
PO Box 446
Hudson, Wisconsin 54016 USA

Library of Congress Control Number: 2006922855

ISBN-13: 978-1-58388-169-9
ISBN-10: 1-58388-169-7

06 07 08 09 10 11 6 5 4 3 2 1

Printed in China

Cover and book design by Dan Perry

Book Proposals

Iconografix is a publishing company specializing in books for transportation enthusiasts. We publish in a number of different areas, including Automobiles, Auto Racing, Buses, Construction Equipment, Emergency Equipment, Farming Equipment, Railroads & Trucks. The Iconografix imprint is constantly growing and expanding into new subject areas.

Authors, editors, and knowledgeable enthusiasts in the field of transportation history are invited to contact the Editorial Department at Iconografix, Inc., PO Box 446, Hudson, WI 54016.

Contents

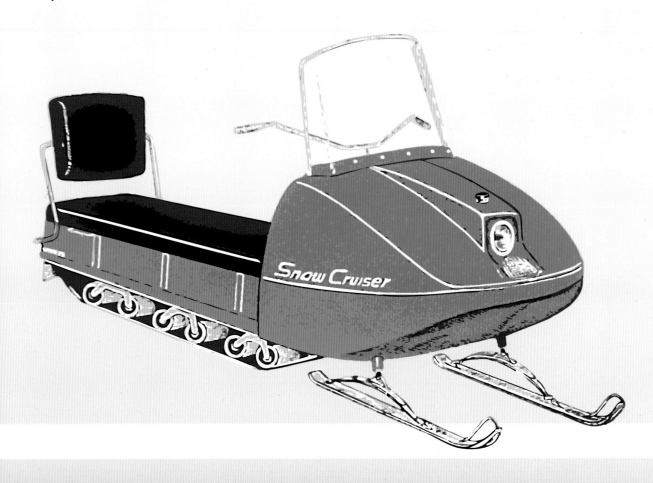

Preface

The Golden Age of snowmobile manufacturing was from 1968-1976. During this timeframe literally hundreds of snowmobile manufacturers sprung up and then disappeared. Some of these manufacturers were huge blue chip corporations like John Deere, to small companies in home garages that most can not even recall their being in existence like a Snow Bug snowmobile.

The collector market for vintage snowmobile items can be easily compared to the interest in vintage motorcycles, farm tractors, gas station items, vintage boats and outboards and similar motorized vehicles. The memories that accompany recreational activities, regardless of the medium, have strong collector followings.

To date there has not been a comprehensive book that specializes in this particular field of interest, with broad appeal, focused on the collectibles that accompanied this incredibly popular recreational activity of snowmobiling.

Vintage Snowmobiles and related collectible items have become one of the fastest growing interest segments in the sport of snowmobiling. Vintage racing is growing every year. The collectibles, or memorabilia, sell for high marks on eBay all year round and at national snowmobile shows. The national publications are now doing segments in each published issue highlighting different aspects of vintage snowmobiles to which I have written several for these publications. Local clubs are sponsoring vintage rides and shows and Baby Boomers and their children are reliving their memories of snowmobiling.

There was a need for a book that would picture many of the collectibles that were produced by the nearly 200-plus manufacturers of snowmobiles that existed from 1968 and into the mid-1970s. By 1984 the list of manufacturers dwindled down to four which still exist today. Collectors new and old are tying to determine what an item is worth and how collectible it is and the current market place did not offer a value reference guide for snowmobile memorabilia.

This book is not all-inclusive. It represents many, but not all, of the different collectibles that were produced during the Golden Age of snowmobiling.

Dedication

This book is dedicated to all those who have suffered with my addiction to snowmobiling over the years.

To my father Dennis Bertolino, for fostering the desire in me to collect and restore old things. For his ability to figure out a way to justify why another old snowmobile needed to be purchased and brought back to life and for his teaching me the ways of paying for all of them. To my mother Kathy Jolin, for her continued patience as a mother listening to her son deliberate for hours on snowmobiling while growing up. In addition to listening, my mother also provided me the moral support and motivation to make dreams happen. To my wife Sara, who can't fully understand my affliction, but accepts it as long as she gets some nice stuff for herself along the way. For her patience and help in traveling around the country to take photographs for this book, and for the assistance in compiling it. To my friends who have been part of my life since growing up in Northern Wisconsin. To Lynn Keillor of Ehlert Publications (*SnowGoer, Snowmobile* and *SnowWeek* magazine), who gave me a chance to dust off my Journalism skills through some freelance writing – I most likely would not have felt I could do this book without having done the freelance work prior. And to Iconografix for seeing the vision I had with this book.

Thank you all.

Credits:

The people, websites and shows below were the ones who made this book possible. Thank you all for your help and assistance.

The Sawyer Family
The Snowmobile Barn Museum
Claude Gendron – Snowmobile manufacturers list.
The Warning Family
A-1 Upholstery Show
Waconia Vintage Show
Jason Remiker
David of www.vintagesnowmobiles.50megs.com
Vintage Eagle River Snowmobile Championship Race.
International Snowmobile Manufacturers Association.

About the Author

I have been actively involved in snowmobiling since I was old enough to stand, literally. (picture)

My parents, Kathy and Dennis, were active in the local snowmobile club and social scene in the Golden Era of snowmobiling in the 1970s. What this really meant was they would be part of a group of friends, who all happened to be teachers at the local grade school. On a Friday or Saturday night, in the cold Northern Wisconsin evenings, they'd ride from one another's homes to the local bars, party into the evening, and then try to ride home on the new snowmachine's they had purchased. Each family had a different brand, and brand loyalty is always strong among friends.

Now, in today's world, this is looked upon negatively, with the high performance snowmobiles that can go over 100 mph on the smooth wide snowmobile trails that zig- zag most of the snow covered states in the 21st century. But in the 1960s and early 1970s snowmobiles did not travel much faster than 30 mph and there were very few, if any marked trails. Not too mention, the reliability was so bad that many times it was not a matter of how fast you got home, but whether you got home on your snowmobile, or on the back of your

friends or spouses. So, the innocence of having fun in the local pubs and driving through the woods and along side the roads with friends was not unusual.

Soon manufacturers took over the sport with over 200 manufacturers sprouting up out of just about every industrial factory around. Watching all this snowmobile activity bustling around me in Northern Wisconsin fostered a longing to be a participant at an early age. I personally can remember being six or seven years old and driving the 1970 Evinrude Skeeter snowmobile that my dad owned for the first time around a field one cold Saturday in Hazelhurst, Wisconsin. My dad assured me that he would be right behind me on the seat at all times in case I got scared. After a while I looked behind to get that reassuring smile of my dad looking at me, but to my horror he was standing about 1000 feet away waving at me. He had jumped off and let me go on my own. Well, the next several years I drove myself around and around the house until I reached 12 years of age.

Upon turning 12 years old, I was finally able to attend Snowmobile Safety Certification Class. Upon completing the required classroom training then getting certified by the WI DNR to drive a snowmobile, I promptly began using my 1974 Arctic Cat El Tigre as my primary mode of transportation. I would travel as much as 20 plus miles in the cold frigid mornings to grade school in Minocqua, Wisconsin, which is about two hours south of Lake Superior. My dad, being the art teacher at my school, would continually egg my friends and peers on by stating that his vintage 1972 Arctic Cat 292 could beat their dad's 1981 brand new Yamaha SS 440s or their own new Yamaha Enticers. How I made it out of school I still can't figure out because he never did race them.

My father was instrumental in my desire to tinker with old snowmobiles. My father was obsessed with finding old snowmobiles that needed work, fixing them up and then selling them. From the early Polaris Sno-Travelers, Ski-Doo's, Arctic Cat's, Sno-Bunny's, and countless others, the time spent slaving for my dad restoring these sleds would only fuel my desires for my own snowmobile. From age 12-18 years old I owned five different snowmobiles all paid for by working for my dad, birthday money and eventually summer jobs in high school.

Luckily enough, I was able to marry a woman who also was brought up with an active family history of snowmobiling. My wife, Sara, was chosen in 2000 by *Snow-Goer* magazine to be a test rider for all the new snowmobiles that season along with three other women. Her opportunity opened the door for me to meet one of the assistant editors of Ehlhart Publications, which is the largest publisher of snowmobile periodicals in North America. Through this relationship developed by Sara, I was able to use my Journalism degree and freelance four articles about different snowmobile shows, swap meets and collectible snowmobiles that I own. Two of the articles were focused on my searching of vintage snowmobile items.

I have also been actively involved with other collectibles over the past 16 years and I have seen how different collectible fields start with just small mailing list, small groups of friends, small club events held locally, and eventually growing to published books on the collectibles accompanied by price list and national conventions. The vintage snowmobile collectible market has just begun to grow from select interest groups, to national organizations and grass roots shows. What has been lacking was a guide to help collectors and enthusiasts realize what is available and what it may be worth to the right person. This is why I have chosen to write this book.

Do you have vintage snowmobile collectibles?

If you have a collection of vintage snowmobile items that you sell or would like to have taken into consideration for inclusion in future reprints of this book you can contact the author Jon Bertolino at:

Jon Bertolino
15N793 Pheasant Fields Lane
Hampshire, IL 60140
Email: Toizrit@comcast.net

Introduction

As is the case with anything that is mass-produced, there usually is a choice in manufacturers. More than likely, you or someone you know, will own a different brand of some item than you also own. At some point, you or they, will probably make a point to tell the other party that yours is better. This difference, whether it be a friendly jab or continual harassment relating to whose car, lawn mower, tractor, snow blower, bicycle, roto-tiller, motorcycle, wave runner (etc.) is better, is inevitable. This difference begins the identification with a particular model and strengthens to brand loyalty. People like to express themselves as being different from others.

Competition on the race tracks all across the regions also fuels brand loyalty for a particular product regardless to what product is being raced. In this case, the snowmobile manufacturers were quick to monopolize on this idea. Brand identification on the race track and snowmobile trails led to brand awareness which increased sales. Even today, with only four manufacturers providing snowmobiles, you can see the brand loyalty everywhere you look. Blue for Yamaha, green for Arctic Cat, yellow for Ski-Doo and red for Polaris. I have yet to see another industry that is so prolific with brand identification. I have seen people nearly get into fights about their particular brand!

The International Snowmobile Manufacturers Association list 1969 through 1975 as the seven highest sales years for snowmobiles to date. For several of these years, sales were twice as large as today's yearly sales. In seven short years the industry went from, for all practical purposes, a hobby industry, to nearly three million snowmobiles being sold by many of the nation's top industrial giants. Snowmobiling had reached a near epidemic proportion in the late 1960s and early 1970s with nearly 200 manufacturers all trying to catch the agile consumer. Brand identification became intense.

Manufacturers put their names and colors on nearly everything they made or could sell the rights to. Included in this book, you will find chapters on many of the most popular segments of brand promotion. Vintage advertisements, containers for oil and fuel, brand specific apparel, vintage toys, manufacturer's signs and shop related items posters, magazines, paper items and lots of miscellaneous items. All branded to help sell snowmobiles or products for snowmobiles.

I am sure you will see many items that you remember using during this era and will wonder what ever happened to those old things? How did we ever get by using them, and I can't believe that they are worth so much now!

Vintage Snowmobile Show.

Vintage snowmobile racing today has a huge following and is getting larger every year.

Photo's from Eagle River, Wisconsin, vintage race series.

Chapter 1

Snowmobilia!

Pricing

The vintage snowmobile collectibles market has been increasing every year for the past three to four years. Attendance at local vintage snowmobile shows has been increasing annually. Entries at vintage races are outpacing entries for late model racing. Competition on eBay is rabid for highly desirable and rare snowmobiles, collectibles and parts.

I have been actively involved with collecting vintage items since the late 1980's. By the time the mid 1990's hit I had fully incorporated a company that specialized in the purchase and sale of many different vintage items from old toys, watches, glassware, writing instruments and eventually vintage snowmobile collectibles. My past 17 years of collecting vintage snowmobile memorabilia through attending flea markets, estates sales, auctions, vintage snowmobile events and extensive monitoring of eBay has given me significant experience to apply towards the current market pricing for snowmobile collectibles.

Pricing is listed for items from a low to high price point based on condition and should only be used as a guide. Many collectibles are not priced and one can project similar pricing as other items that are in the same chapter. Some items are not priced based on the values not being substantially notable. Pricing is subjective. Many items listed in this book could sell for, or have sold for significantly less or more. The price range contained within is to be used as a guide to help assist the buyer, or the seller, in their establishing a baseline to what the worth of the particular item is. Sellers who offer items depicted in this book for sale at a higher pricing may be applying their own assessment of value of what an item is worth as opposed to trying to take advantage of buyers. Some items may also appreciate faster than others. Remember, the item is only worth what you are willing to pay for it.

How to Find Vintage Snowmobile Collectibles

I know you are a vintage snowmobile fan or you would not have purchased this collectors pictorial guide. I also have a good hunch that by the time you finish looking at all the different collectibles that are pictured within this book you will be much more aware of the hundreds of different snowmobile collectibles that were produced from 1968-1983 and most likely have a rekindled desire to seek out some of these items.

Now, the big question is "how do I find these items?"

I will try to explain some of the best methods to acquire snowmobile collectibles along with areas to find them. All of which should help you in your search for snowmobile collectibles and many times the sleds that made them possible. We will start off with the best areas to search out snowmobile collectibles.

1. Garage and Estate Sales:

Garage and estate sales offer probably the best venue for finding vintage snowmobile items and many times snowmobiles themselves. Why? Well, first for starters: you are in the person's garage, basement, house, yard, barn, or shed. And what usually resides inside these buildings? Old snowmobiles. Usually you can ask the host of the garage sale whether or not the old snowmobile is for sale, or if the snowmobile is not for sale, do they have any old clothing, helmets, oil cans etc. You will be surprised what these people can dig up. I have found many old snowmobile helmets, snow suits and related items many times for only a couple dollars. This past year I found a 1976 Yamaha SRX 340 sitting in a barn in nearly mint condition one mile from my house. Again, the price was significantly lower than it was worth and sold for significantly more shortly thereafter.

2. Automotive and Related Swap Meets:

Surprisingly enough, one of the best places to find snowmobile collectibles is automotive or motorcycle swap meets. Just about every state has some really big

swap meets for old cars and motorcycles. Many of these are listed in different automotive magazines and newspapers. The reason that you will find snowmobile items at these shows is because the collectors of old cars and bikes many times are buying out old barns worth of cars and car parts, or buying old filling station items. The snowmobile items are located in the garage or storage building most of the time along with the automotive items. What usually happens is that the sellers of these items want the buyers to: "take it all or nothing." These items then end up at the automotive swap meets and the snowmobile oil cans, helmets, used parts, manuals, etc. are put on the table for sale.

3. Flea Markets:

The local flea market can be a great venue for finding old snowmobile items. Just about every town has a local flea market. Some regions even have large national flea markets that bring in thousands of attendees with 100's of vendors from all over the country. Why are flea markets a great place to find snowmobile collectibles? They are because these dealers get the majority

of all their "junk" from where else? Garage and estate sales. See number one above.

4. Advertisements in Local Papers:

You have to remember, that just because you know in your heart that a pink snowmobile suit with a fuzzy purple collar is worth over $100, you need to remember that most people would not be caught dead wearing one of these vintage snowmobile suits in today's world. So, understanding this, many people would rather throw an item like this, or similar items, in the trash can as opposed to putting it out at a garage sale. Many owners figure who in their right mind would actually want to buy such an ugly snowmobile suit, not to mention that they are not warm. So, if you can not reach these people through their garage sales then you have to try to reach them a different way. I recommend putting "Want Ads." in your local news paper and shoppers. You will be surprised how many people will call you stating that they have a bunch of old items from when they were into the sport in the 70's. Plus, there is very good chance you could land some nice vintage snowmobiles for very little money.

But remember, once you begin advertising for these items, people will then assume that what they have is worth something. Why else would you be spending your money looking for it! Then others will be just glad to get rid of the stuff!

Below is an example of an advertisement that you could place:

Wanted: Collector looking to purchase old snowmobile items from 1968-83. Interested in helmets, snow suits, oil/gas cans, old parts, literature, posters, and all miscellaneous paraphernalia. Also interested in old vintage snowmobiles. Need not run, will pick up. Cash Paid. xxx-xxx-xxxx

5. Word of Mouth:

My thoughts are that word of mouth can be the least expensive and most rewarding method for finding old snowmobile items. I have personally found a 1978 Polaris RXL Sno-Pro sitting in a brush pile in Michigan. All I did was ask a customer of mine at the end of meeting if he ever did any snowmobiling in this part of Michigan. Turns out he used to race these sleds. I also asked a customer of mine if they get much snow in this part of Indiana. Turned out the guys family used to own an Evinrude Snowmobile dealership in the late 60's and 70's. I ultimately purchased all the remaining dealer parts and inventory they had put away in the attic 15 years ago. Just recently, I was able to purchase one of my favorite Yamaha snowmobiles, which was a 1980 deep red SS440 from my Great Uncle. It has been sitting in his storage shed for nine years with a blown piston. I just asked him at a family gathering if he still had that sled I remembered. Sure enough he did.

6. Snowmobile Shows:

This one is pretty obvious. Probably the ultimate place to find vintage snowmobile items because the entire focus of the show is on snowmobiles with many shows catering to vintage collectors and restorers. Most of the snow shows have swap meets. These shows are just busting open with old snowmobiles for sale. In addition to the sleds that are for sale, you will find many sellers offering some related items such as covers, fuel containers, old videos, posters, accessories and more. But remember, you are at a show that specializes in this

merchandise. Many of the sellers are savvy collectors like you who just happen to have some extra items for sale. If you are looking for specific items for specific brands though, these shows are a must. Many of the dealers specialize in one particular brand and they have the item you are looking for, or know of someone who may. I highly recommend attending some of these shows. To find a listing of vintage snowmobile shows in your area you can contact several organizations. The VSCA or Vintage Snowmobile Club of America and the Antique Snowmobile Club of America both list different shows that are coming up. In addition, there are several websites on the Internet that list the shows. Just type in Vintage Snowmobile Club or Show and you will find plenty of links.

7. Internet:

The biggest thing to hit the field of collecting anything vintage has been the growth of the Internet. By now, most of you will have already bought computers, learned about how to buy online, and realized that you are not the only one out there that has an interest in a particular subject. Just type in the name of something that you enjoy and you will probably find thousands

of related websites that are all about what you are interested in. If you have not begun using the Internet, you better start soon because you are missing out on the fastest growing medium for finding vintage snowmobile collectibles. I recently purchased a complete engine out of Finland for one of my rare 1982 SRX 500's by networking with snowmobile enthusiast on a snowmobile website.

The Internet will allow you to find websites that are dedicated to most of the popular brands of snowmobiles and collector sites. Through these websites you will be able to network with collectors just like you who have the same passion for snowmobiles that you have. In addition, the Internet will allow you to shop eBay which is basically the world's largest garage sale. My take on eBay is this: If you can not find it on eBay, then you have not looked hard enough or long enough. The rare items that I have sold, purchased, and seen sold and purchased on eBay, continues to absolutely amaze me. Words can not explain what you can find and sell on eBay. I have found items for my snowmobile collection located in all corners of the nation and sometimes world for auction on eBay. If you are not monitoring eBay for those particular vintage snowmobile collectibles that you have always wanted, you truly need to, because more and more everyday people have begun to realize that they can find buyers on eBay for just about anything and everything that they have in the garage or basement. They also know that people are willing to pay more for these items than the $1.00 they might ask for at their garage sale.

Vintage Snowmobile Museums

Vintage snowmobile collecting is getting increased interest every year. Vintage snowmobile shows and events are popping up in small towns from the East to the West not only in the winter but throughout the year. It appears that the snowmobile collector just cannot wait until winter to enjoy their sport. As the sport has grown, so has this desire to view vintage snowmobiles year round. If you crave looking at the vintage iron that is out there, or looking at really cool collectibles related to snowmobiling I highly recommend visiting some of the great museums that are available for all to attend. I have listed some of the more popular museums that offer the most snowmobiles and history for viewing. Some are operated by the manufacturers, and others by private collectors, all of which offer great historical references and displays.

The New Hampshire Snowmobile Museum
PO Box 10112
Concord, NH 03301
603-648-2304
Information: info@nhsnowmobilemuseum.com

The Snowmobile Barn Museum
928 Cedar Ridge Rd.
Newton, NJ. 07860
973-383-1708

The Snowmobile Hall of Fame Museum
PO Box 720
St. Germain, WI. 54558
715-542-4477
Information: hallfame@newnorth.net

J. Armand Bombardier Museum
1001 Avenue J.-A.-Bombardier
Valcourt, (Quebec) J0E 2L0
450-532-5300
Information: info@museebombardier.com

Polaris Experience Center
Suite #2
205 5th Ave. SW
Roseau, Minnesota 56751
(218)463-2312

Hall of Fame Museum in St. Germain, Wisconsin.

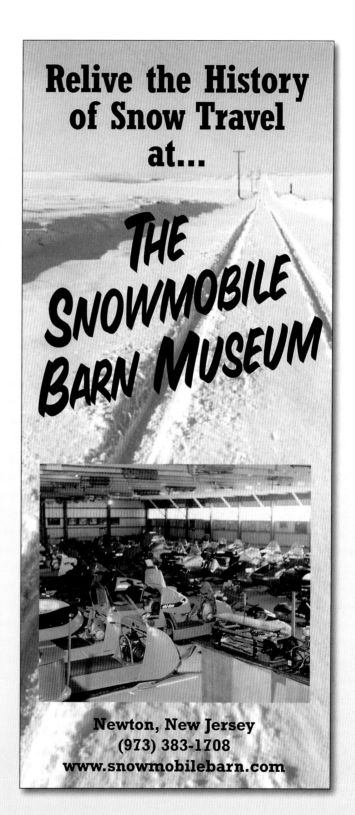

Chapter 2

Vintage Containers

Fuel, Oil, & related

Snowmobiles were everywhere in the early 1960s and 1970s. Hundreds of brands all jockeying to be the market leader. Each with their distinctive features and benefits, most with their matching attire and accessories. One thing that all the manufacturers had in common was the common denominator of gas and oil.

Most of the largest manufacturers had their own branded, and sometimes colored, oils to match their sleds. Some of the most valuable fuel and oil containers today are the vintage models that represented each manufacturer's particular brand. Collectors many times will pay premiums to get vintage containers that match their favorite snowmobile. As you can imagine, with many of these vintage snowmobile brands only being in business for a couple of years, the ease of finding branded oil and other lubricant containers can be quite a challenge.

The major, and not so major petroleum companies, jumped at the opportunity to brand their oil or gas as the best for 2-cycle snowmobiles. They saw a new emerging market for oil, gas, and lubricants presenting itself and were not going to be left behind. You will see oil cans in this section from hundreds of companies. The ironic point is most of the oil was produced and then branded and dyed by about three to four companies who supplied the whole industry!

This chapter is one of the largest in this book because this segment of the sport offers so many varieties for the collector to acquire. Not only is there a large volume of branded containers to find, but the containers can be easily displayed in the garage or collector's room which makes collecting that much easier. Like any collectible, if it can be easily displayed it is more likely to be sought after. Not everyone can store a hundred vintage snowmobiles. But just about everyone has room for a hundred vintage oil cans representing a hundred different manufacturers!

What makes some containers highly sought after is not just the manufacturer's names but the artwork applied to the container. Containers that are bright with high quality artwork are the most sought after containers regardless of the manufacturer. On average the collector can expect vintage snowmobile cans to be valued between $10 to $25 a container with premiums on top of that for high graphic containers and manufacturer specific containers. I have put some notes next to specific containers that draw above-average prices.

Fuel and Oil Containers: Five Gallon Cans

Description: Five gallon gas cans are hard to find and are very collectible based on their display qualities. Premiums are paid for snowmobile manufacturers cans when compared to general gas or oil companies' cans. Dents and rust detract from values.

Arctic Cat gas cans. $50-$100

Polaris gas can. Plastic. $20-$40

Miscellaneous oil/gas manufacturer's cans. $20-$40

Bardahl, very hard to find. $30-$50

Most common gas can.

Miscellaneous five-gallon gas cans. $20-$50

Very common.

Fuel and Oil Containers

This section encompasses many of the different oil containers that were produced during the Golden Years. Premium values are paid for oil containers that were produced by popular manufacturers or ones that depict interesting artwork. Most containers can be valued between $10-$20 with some reaching over $50.

Allied Leisure Products. Pictures vintage Polaris TX. $20-$30

Very early Arctic Cat oil can. Front and back pictured. Very rare. $50-$150

Arctic Cat Purple Power Lube. Several different versions. Tall container is a Canadian issue. Highly collectible. $15-$40.

Autoboggan. Very rare. $20-$40

Bolens Sprint Oil. Great artwork and a manufacturer's oil. $20-$40

Bardahl. Very easy to find with great artwork. $5-$15

Older version. $15-$20

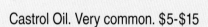

Castrol Oil. Very common. $5-$15

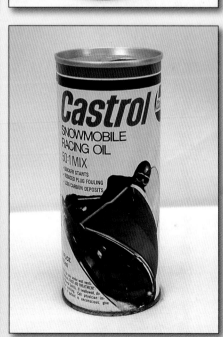

Co-Op. Nice colorful artwork. $10-$20.

Drag-On. Neat artwork. $10-$20.

Flash 401 Oil with premium hat offer. $25-$50 with hat, $10-$15 without hat.

23

Golden Spectro. Great artwork. $20-$30

Pictures a Rupp snowmobile.

Harley Davidson snowmobile oil. Very hard to find. Also collected by motorcycle collectors. $25-$50

Holiday gas station snowmobile oil. Beautiful artwork. Highly collectible. $30-$60

Hudson Bay. Good snowmobile artwork. $15-$25

Itasca. Attractive artwork. $20-$35

John Deere snowmobile oil. $10-$20

25

Kawasaki snowmobile oil. Very hard to find. $15-$25

Kendall. Great artwork. $15-$25

Klotz racing oil. High demand for this company's oil cans. $20-$40

Motomaster. Colorful artwork. $20-$30

Kohler. Very rare. $30-$50   Mercury snowmobile oil. Hard to find. $15-$25.

Moto-Ski oil display from a dealer show room. Rare. $150-$300

Oilzum. Great artwork. $10-$20

Polaris. Very rare plastic snowmobile oil bottle. $50-$100

Penn. Great artwork. $20-$35

Rislone. Commonly found but very colorful with attractive artwork. $15-$30

Yamaha. Manufacturer's oil cans. Hard to find. $25-$50

Shell. Very colorful with good artwork. $15-$30

Sanko. Attractive artwork. $15-$25.

Scorpion. Manufacturer's oil can with display. Very rare. $75-$150

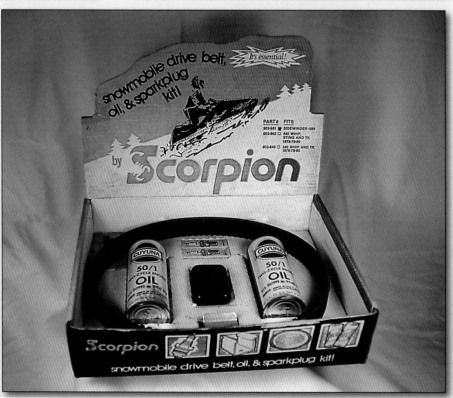

Ski-Doo oil cans. $15-$30. Higher value for metal cans compared to plastic.

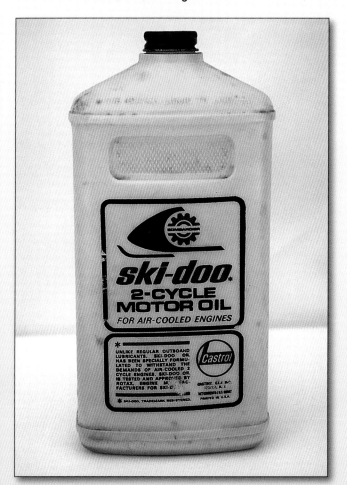 

Castrol branded – Ski-Doo oil can. Very rare. $25-$50

Snobil. Popular container very colorful. $10-$25

Very popular. $20-$40

Sno Oil. Nice artwork. $15-25

Viking. Manufacturer's oil cans. Very collectible. $20-$40

Wynn's oil. Very popular. $10-$25

Polaris. Artwork of Polaris TX. High demand. $20-30

Polaris manufacturer's cans. $20-$30

Rupp. Manufacturer's oil can. Very rare. $20-$40

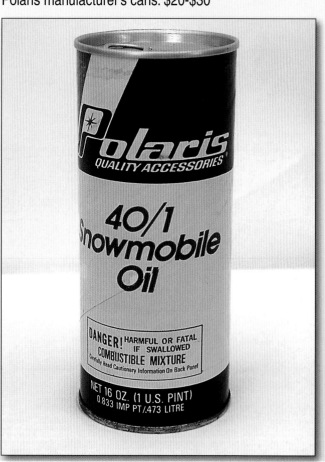

Arctic Cat manufacturer's oil cans. Highly collectible. $15-$30

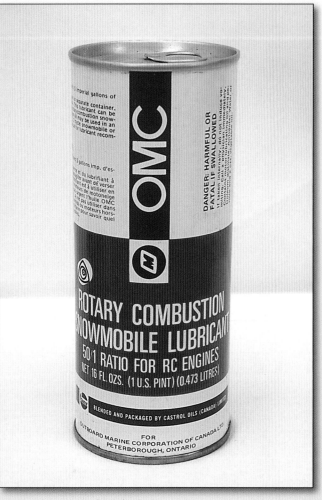

Johnson/Evinrude oil containers, fairly common. $10-$15

Scorpion. $15-$25

Scorpion manufacturer's oil can. $25-$40

Although plastic, the Shell oil containers exhibit excellent artwork. $20-$35

Major oil companies. $10-$15

Very popular oil container. How exactly is Analube pronounced? $10-$15

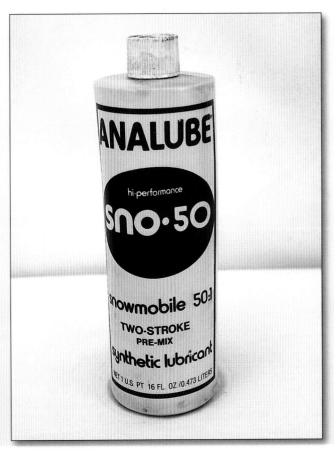

Nice artwork. $15-$20

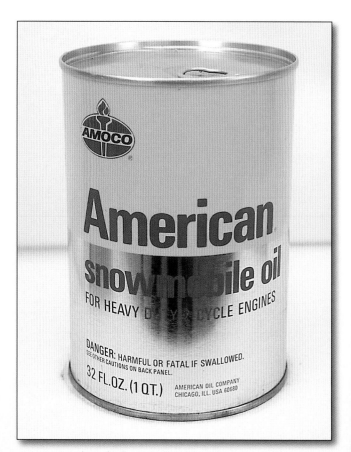

Very plain. $10-$15

Very plain. $10-$15

Snowflakes add to desirability. $15-$25

Excellent artwork for plastic. $15-$30

Very plain. $10-$15

Nice artwork. $15-$20

Very plain. $10-$15

Very plain. Union Carbide made many of the snowmobile oils for other companies. $10-$15

Nice artwork but not related to snowmobiles. $10-$15

Cuyuna is also Scorpion oil. $15-$20

Stihl, known for their chainsaws, not snowmobiles. $10-$20

Great oil advertising thermometer. $50-$75

Nice artwork. $15-$20

Very common. $10-$15

Excellent artwork. $20-$30

Very plain. $10-$15

Very plain. $10-$15

Some light artwork. $15-$20

Good artwork. $15-$25

Excellent artwork. $25-$35

Hard to find but very plain. $10-$15

$10-$15

$10-$15

Attractive oil can. $10-$20

$10-$15

Quite easy to find can. $10-$15

Simple but attractive. $15-$20

Nice artwork. $10-$20

Attractive oil can artwork. $15-$25

Plain artwork, low value. $10-$15

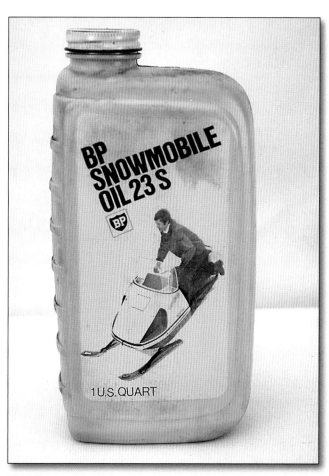

Excellent artwork for plastic container – pictures a Ski-Doo. $20-$40

Very attractive artwork. $20-$30

Nice artwork. $15-$25

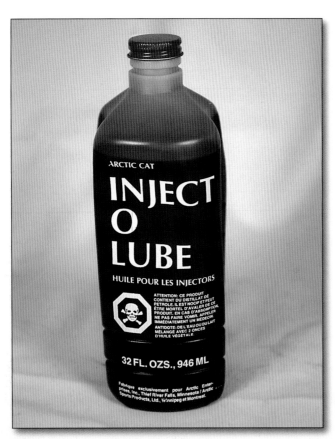

Relatively hard to find Arctic Cat injector oil, poor graphics. $10-$15

Both oil cans depict nice artwork. $15-$25

Attractive shape and artwork. $15-$30

Simple artwork and plastic. $10-$15

Very nice artwork on both of these containers. $20-$30

Miscellaneous Lubricants and Paints

Many collectors try to source out all shop repair items which can include branded miscellaneous lubricants and OEM paints. These items all make for great display items for the vintage collection. Many are very hard to find because they were limited use items. Most items range between $10 and $30 and all are from the 1970s.

Great assortments of vintage OEM spray paints. Many different brands are represented.

Very old spray cans. The Auto-ski is very rare.

Very hard to find OEM quarts of oil based paint. $20-$30

$5.00 value.

Very rare. $10-$20 each

Clutch lubricants. The Sno-Jet is especially collectible.
$10-$20

Excellent artwork on these two. $10-$20 each

Manufacturer's supplies. $10-$20 each

Manufacturer's supplies. $10-$20 each

Different aftermarket lubricants. $5

Premium value for artwork. $15-$20

Spectacular artwork. $20-$30

$5.00 value

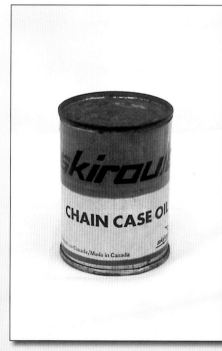

Very hard to find manufacturer's chain case lube cans, unopened. $10-$20

Chapter 3

Vintage Apparel

Vintage apparel is by far one of the wildest things to think of as a collectible. What is so wild about vintage apparel? Well, it is hard to imagine that people wore these items regularly and were actually making a style statement in conjunction with exhibiting brand loyalty. Another reason why it is such a hot collectible is that the ever-increasing vintage snowmobile market is drawing demand on these clothing items. Many die-hard collectors of vintage snowmobiles want to have as much of the related merchandise as they can find for their favorite brand or particular sled. This demand has made prices for select items increase dramatically.

This chapter will picture interesting clothing related items. You will see helmets in bright colors with logos, wildly colored snow suits, patches, boots, hats, and just about anything that could be worn to either keep warm or make a statement about the brand of snowmobile you drove.

One particular field of interest from this era were the sewn on patches that were the craze of the Golden Years of snowmobiling. Patches were extremely popular. This, combined with the sport of snowmobiling being club driven and a brand focused sport, led to hundreds of club or brand specific patches being produced. Some of these patches will put a real smile on your face. You can almost see the cold snow suited parent or kid sitting on his snowmobile with his or her snowmobile suit full of these things!

An example of appreciating apparel values would be vintage helmets and snow suits which can sell for hundreds of dollars over what they cost new. Just because it is old, ugly and outdated, and you think: Why would anyone wear this old piece of clothing? Doesn't mean that there is not a collector out there just waiting to find one.

Brand Specific Apparel

Vintage apparel is very hot as a collector's item. Although some people who are not vintage snowmobile enthusiasts may find the colors and designs to be very hard on the eyes and dated, collectors find that the wilder the vintage clothing looks, the more desirable it is. It makes a statement of the era and brand.

Arctic Cat apparel

1973-1975 Jacket. $30-$50

1972 Snowsuit. $25-$40

Late 1960's Arctic Cat Snowsuit. Premium value with spotted cat pattern hood. $50-$85

Sweaters. $20-$40

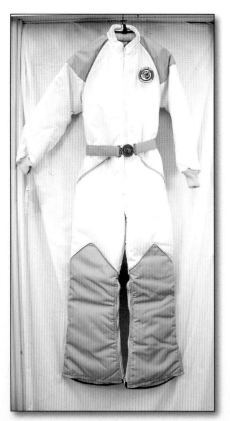

Ladies 1972-1979 snowsuits. $50-$75

1972 Furry Arctic Cat jacket. Very collectible. $90-$150.

1972 windbreaker summer jacket. $30-$50

Men's and Women's
1973-1979 snowsuits.
$30-$50

1973-74
Arctic
Cat
ladies
purple
snow
suit. $50-
$100

1980
Arctic
Cat
snow
suit.
$50-$85

Moto-Ski apparel

1970s Moto-Ski snowsuits and jackets. $30-$50

Lightweight shop or summer jackets. $30-$50

1970s Ski-Doo jackets. $30-$40

1974-1975 Ski-Doo TNT jacket. $30-$40

1970s Ski-Doo snow suits. $50-$75

Ski-Doo jacket. $30-$40

Ski-Doo snow suit and matching gloves. $50-$100

Ski-Doo snow suits. $50-$75

Ski-Doo sweaters. $20-$40

Yamaha apparel

Yamaha jackets. $40-$75. Matching snow pants, add $25.

Yamaha SRX leather jacket. $75-$125

Yamaha 1979-1981 SRX sweater. $30-$40

1981 Yamaha SRV and SS snowmobile sweater. $20-$35

1970s Yamaha jacket. $40-$50

1980-1981 Yamaha SRX snow suit. $50-$100

John Deere sweater. $20-$40

Chaparral jacket. $40-$60

Mercury Trail Twister Jacket. Very hard to find. $75-$100

Skiroule snow suit. Very rare. $50-$100

Polaris jacket. $20-$40

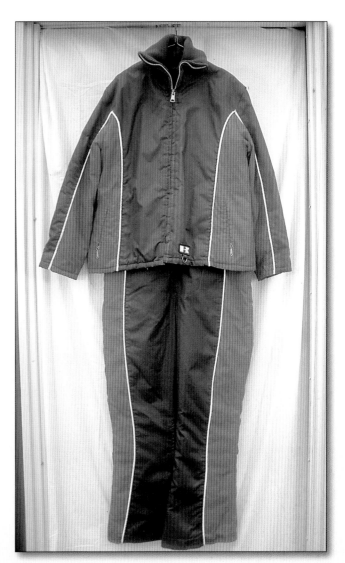

Polaris snowsuit. $50-$75

1973-1975 Harley Davidson Snowmobile Jacket. Very collectible. $50-$75

Sno-Jet snow suit. Very collectible. $75-$100

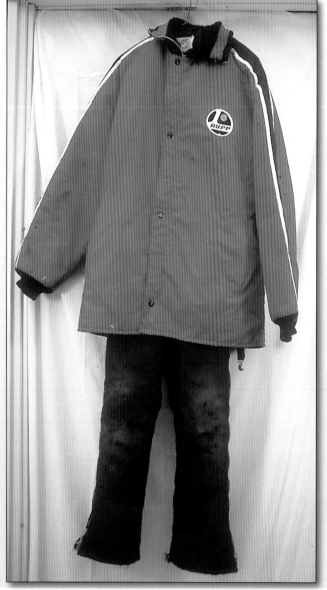

Rupp snow suits and jackets. $50-$75

Miscellaneous Apparel Items

Boots, hats, face mask, and mittens were sold as accessories. Many of these items can be easily found but items associated with brands long out of business are true finds.

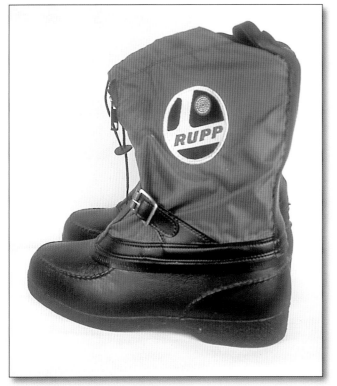

Rupp Boots. Very rare: $20-$50

Ski Whiz boots. Very rare: $30-$75

Polaris boots. $20-$50

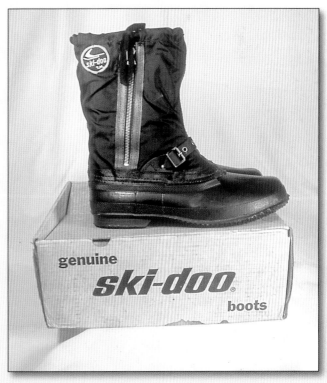

Ski-Doo boots. $30-$40

Chaparral boots. Very rare. $30-$75

Arctic Cat boots from 1972. $20-$40

Moto-Ski boots. $20-$40

1971 Arctic Cat painter's cap. $30-$50

1972 Arctic Cat fuzzy cap, three colors. $30-$50

1974-77 Arctic Cat face masks. $10-$25

Mid 1970's Arctic Cat multi colored fuzzy gloves. $35-$50

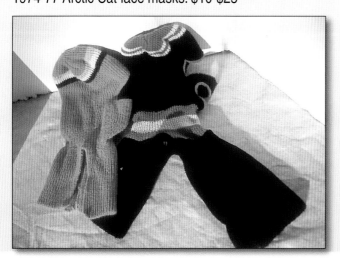

Late 1970's Yamaha racing cap. $30-$50

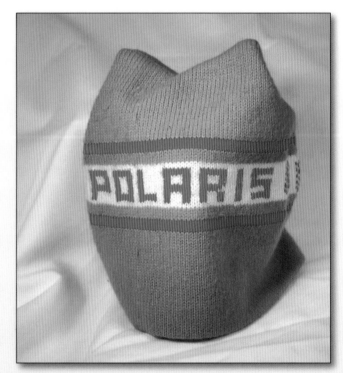

Polaris hat. $10-$25

Early 1980's Yamaha hats. $10-$25

Sno-Jet and Kawasaki knit caps. Rare. $20-$35

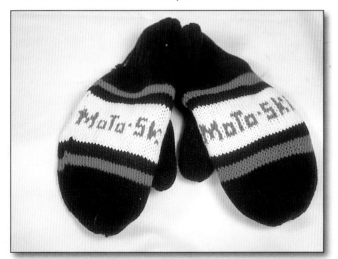

1970s Moto-Ski mittens. $10-$20

Shop hats. $10-$20

Ski-Doo and Moto-Ski knit hats. $10-$20

Vintage Helmets

Manufacturers wanted their colors on everything and helmets were not going to be forgotten. They are a very visible portion of a snowmobiler's attire. Some helmets command a premium based on unique colors and designs or for the specific year of production that coincides with popular snowmobile models. Many helmets sell for $30-$75.

Arctic Cat purple metal-flake racing "horned" which means the side bulges on the helmet. Circa 1970-1972. $50-$100

No "horns" or "bulges" on sides. $30-$40

Arctic Cat helmets. $25-$75

Arctic Cat Pantera Snowmobile helmet, circa 1978. $20-$35

Arctic Cat Snowmobile helmet, circa 1976. $35-$60

Polaris helmets full face models. $40-$75

1977 Polaris Tx helmet. $40-$100

1976-78 Polaris full face helmet. Metallic colors. Mint in Box example. $50-$150

Late 1970's Scorpion helmet. $30-$50

1976-78 Polaris full face helmet, three solid colors. $50-$120

Premium values for helmet in original box, add 30% to value.

Moto-Ski helmets. $30-$40

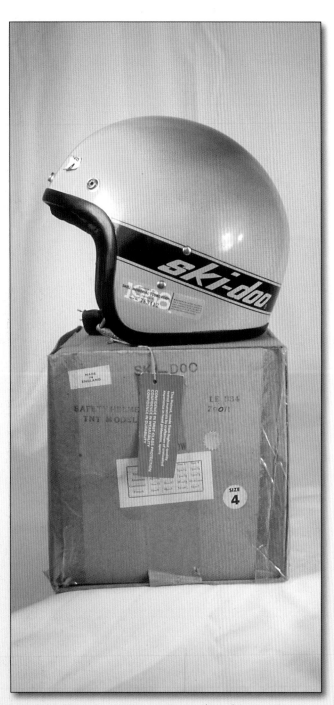

Very old Ski-Doo helmet with box. $50-$75

Vintage Ski-Doo helmets. Premium values when specific model names are embossed on helmets. $50-$75

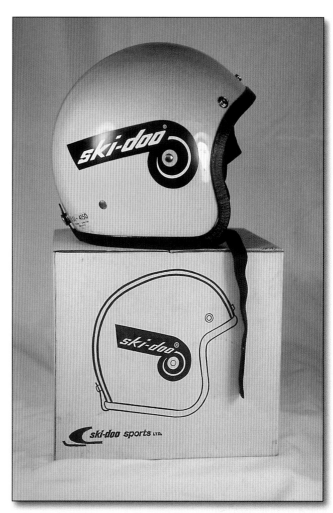

Various 1960s-1970s helmets. $40-$50

Yamaha 1970s helmets. Very hard to find.
$85-$125

John Deere helmet from early '70s. $30-$50

John Deere helmet. Very collectible full face version. $40-$75

Scorpion Helmet, circa 1974-1976. $50-$80

Scorpion Helmet, circa 1979-1981. $50-$100

Vintage Patches, Stickers, and Pins

During the golden age of snowmobiling, club membership was as important to the snowmobilers as the sleds they rode. Almost every snowmobiler had club patches or other patches on their snowsuits advertising their club or favorite brand. Many times it was less expensive to buy a brand specific patch and put it on a generic snowsuit. You will find many patches in this section for many different brands.

Typical patch adorned winter vest.

Sno-Jet promotional pin. $5-$15

Miscellaneous brand patches. $3-$10

Club patches. $3-$10

Club patches.
$3-$10

Miscella-
neous brand
patches. $3-
$10

Arctic Cat patch-
es. $3-$10

Moto-Ski and
Yamaha patches.
$3-$10

Miscellaneous patches.
$3-$10

Rupp patches.
$3-$10

Ski-Doo patches. $3-$10

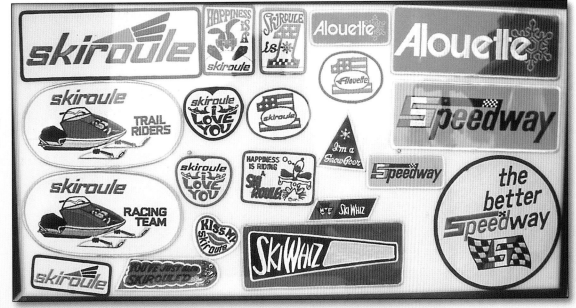

Miscellaneous brand patches. $3-$10

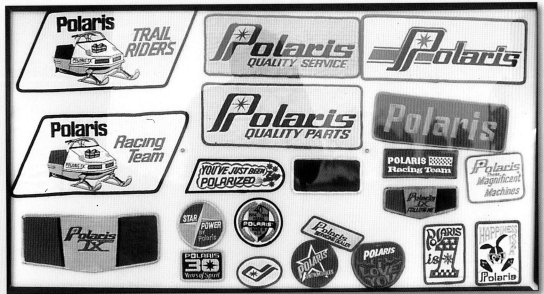

Polaris patches. $3-$10

Ski-Doo and Moto-Ski patches and stickers. $3-$10

Snowmobile advertising pins. $5-$15

Miscellaneous brand patches and stickers. $3-$10

Miscellaneous patches and stickers. $3-$10

Arctic Cat stickers. $3-$10

Chapter 4
Vintage Toys

This is the chapter that started my personal collection of vintage snowmobile collectibles. From the beginning I can remember playing with several Polaris plastic rubber-band wind-up snowmobile toys which were about the size of a Matchbox toy car. I think I must have snowmobiled every square inch of carpet on those little plastic toys when I was in 1st and 2nd grade. From my memory, I can remember getting the toys in cereal boxes. But I have recently found out that you could get them at Polaris dealers in the early 1970s too. I can also vividly remember playing with my friend's 1973 Tonka Arctic Cat El Tigre and Ski-Doo Silver Bullet. Both were friction powered—just push backwards and they zoomed forward on the hardwood floor. However, they were much more fun to be played with outside in the snow.

The Holy Grail of snowmobile toys was the coveted Arctic Cat Kitty Cat miniature gas powered snowmobile. (See accompanying picture.) This snowmobile is not quite a toy, but to an adult it is. To a child growing up around snowmobiles this was the absolute dream. I can not begin to tell you how many times my friends and I begged, pleaded and justified to our parents to why we needed to have one of those Kitty Cats. Quite frankly, I still want to own one of these sleds just to fill that void that never got filled as a child! My childhood friends and I still talk about the fun we had with those little snowmobile toys.

As you can see the goal of the snowmobile manufacturers was met through marketing toys and games to children. There probably is not a better way to get brand awareness to future buyers than have future snowmobile buyer's play with your branded toys as children. The concept is still in practice today. Most of the remaining four manufacturers market toys, or miniature snowmobiles, to children.

In this chapter you will see many of the different toys that were promoted either by the manufacturers themselves or by the many different toy manufacturers who wanted to cash in the opportunity to sell stuff to this exploding market place.

Vintage snowmobile toys and collectibles are quite difficult to find based on the toys having been played with extensively and then discarded after they were broken, lost, or given away. Once the children grew up, there was not a reason to keep the items around.

1973 El Tigre Tonka friction toy, mint in box. $50-$150

Scarce 1974 Arctic Cat El Tigre Tonka friction powered toy. $50-$125

1973 Ski-Doo Silver Bullet Tonka friction toy. $50-125

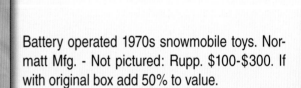

Battery operated 1970s snowmobile toys. Normatt Mfg. - Not pictured: Rupp. $100-$300. If with original box add 50% to value.

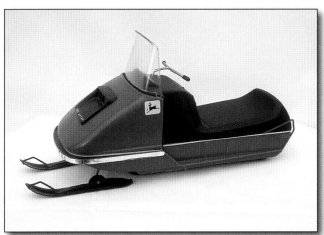

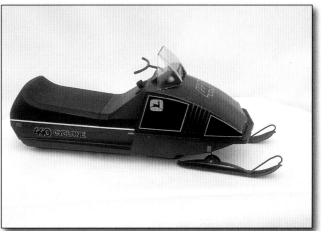

Battery operated 1970s snowmobile toys. Normatt Mfg. - Not pictured: Rupp. $100-$300. If with original box add 50% to value.

SnowHawk. Mfg. Marx. Inc. $50-$100

Snowmobile model kits. Circa 1970s. $20-$75

Very popular Arctic Cat el Tigrè with color-matched truck. $50-$75

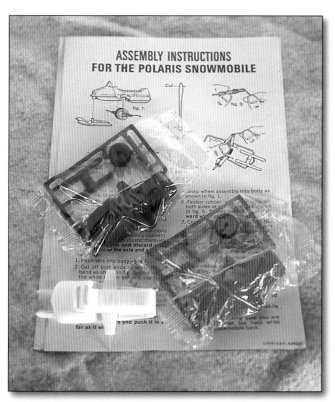

Gas powered Cox Mfg. toy, 1973 Ski-Doo Silver Bullet. $100 loose, $400 mint in box.

Polaris rubber band powered plastic toys thought to have been given as a premium in cereal boxes and at Polaris dealers. $10-$25 loose, $35-$40 in package.

Tonka truck and trailer with snowmobile. $30-$40

Tootsie Toy Mfg. Die cast toy car and trailer with snow-mobile. Circa 1969-1973. $20-$50

1970s mint in box. $20-$40

Large nine-piece set with snowmobile and artwork on box. $50-$85

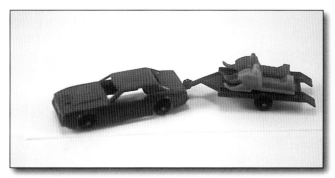

1970's Midgetoy Mfg. -Rockford, Illinois. Two snowmobiles, trailer and car. $20-$35

Plastic toys by various manufacturers. Circa 1969-1975. $10-$20

Fisher-Price toy, common. $10-$20

Toy by Mego. $20-$30

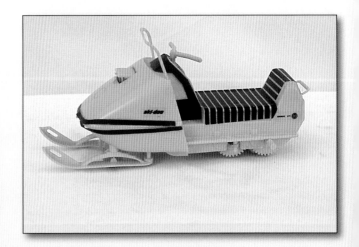

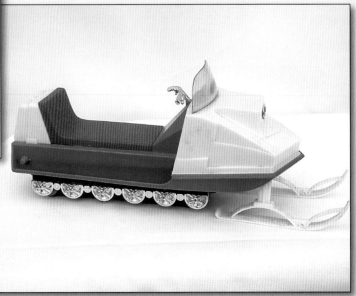

Plastic toys by various manufacturers. Circa 1969-
1975. $10-$20

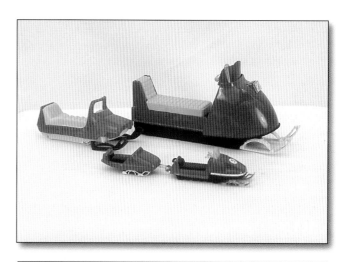

Miscellaneous plastic snowmobile toys. Some are by unknown manufacturers and many look similar to Ski-Doo Snowmobiles. $15-25

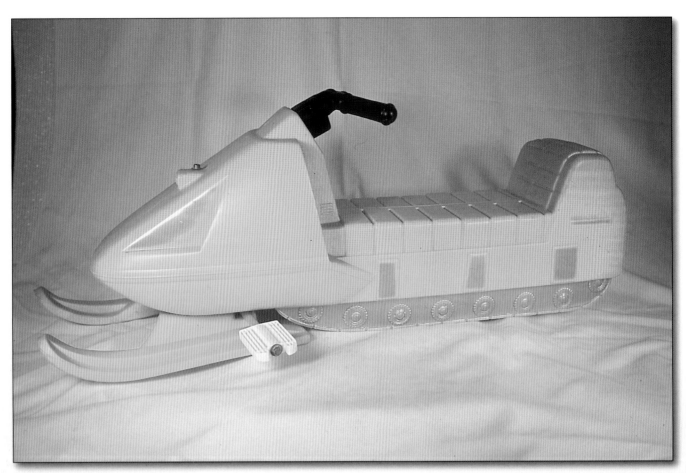

Plastic sit-on snowmobile pull toy.

Very rare Aloutte model. $50-$150

Vintage board games. All are very difficult to find. $50-$150

Scorpion game with excellent artwork.

Ski-Doo game with great artwork.

Moto-Ski game.

Beat the Cat board game, circa 1971, with Arctic Cat pictured. $40-$60

Ski-Doo transfer stickers. $10-$20

Aurora slot car snowmobile. $30-$50

1970s Ski-Doo puzzle. $10-$20

Great artwork. $25-$40

Small diecast toys. $15-$25

Vintage lunch box and thermos. Very difficult to find. $50-$125

Bandai Made in Japan. Battery operated steel toy snowmobile with remote control. 12" x 8" box. Very unique and hard to find, circa 1970. $100-$150

1972 Arctic Cat Inflatable House Kat air filled toy. $85-$150

Chapter 5

Vintage Signs
and Dealer Related Items

This chapter encompasses one the hottest segments of vintage snowmobile collectibles. As collecting vintage snowmobiles continues to expand with grown up kids and older adults trying to relive the glory days and memories they had in the late 1960s through early 1980s, the demand will have far outpaced the supply for their favorite manufacturer's store, or dealer specific collectibles. Collectors very much pride themselves in being able to have dealer-only collectibles and especially the large lighted signs glowing in their display area over their restored vintage snowmobiles. Just like with vintage car and gas station collectibles, the advertising signs are very hot.

This chapter really focuses on the extremely hard to find dealer-only branded items that you could only find at a manufacturer's dealer. For example: your local snowmobile dealer usually had a dealer sign that advertised the specific brand of snowmobile that they sold which would be hanging outside the storeroom or attached to the building. Once inside the dealer's showroom a person usually would have seen dealer-only posters on the wall promoting new snowmobiles by the manufacturer, dealer clocks, hanging mobiles and some point-of-purchase displays.

What makes these items so incredibly rare is that for starters most of these items were advertising the new products that were offered that year. Hence, the next year these items were obsolete and discarded. This is the case with most paper items like posters and point-of-purchase displays. The outside dealer signs are so difficult to find because many of the brands were in business for only a few years and dealers usually did not hang signs outside to advertise a brand that was not in business anymore. Also, existing manufacturers usually updated their look and colors every so many years in order to keep a fresh look, so down went the old and up went the new.

Collectors prize posters because they are so easy to hang in the garage with their display of vintage snowmobiles. Dealer posters are again very challenging to find because they were truly throw-away items and represented a specific year. Most posters can fetch between $40-$100 when in good condition and some even more.

Many of these outdoor lighted or non-lighted signs can sell for between $500 and $1,500, and sometimes even more. If you are fortunate, you may still be able to find some of these signs sitting behind old dealers or maybe even still hanging.

Snowmobile Manufacturers Dealer Signs

Description: Dealer signs were available as lighted or non-lighted. They often were hung outside the dealers or along highways. Demand is very high for popular and current manufacturer's signs. But companies that are no longer in business are truly a rare find. Premiums are paid for large lighted glass signs that still work. Cracked glass or excessive rust reduces value. Most dealer signs are valued between $200-$800

Lighted Alouette sign. $500-$800

Alouette Snowmobile dealer sign, circa 1970s value. $200

Very rare 1960s Arctic Cat lighted sign. $600-$1,500

Metal Arctic Cat sign. $250-$400

Common early 1980s dealer sign. $250

1971-72 Arctic Cat Metal Dealer Sign. $400-$800

Arctic Cat lighted sign. $400-$800

Very rare Boa Ski signs. $500-$800

Small Chaparral lighted sign.
$200-$350

Very hard to find.
$500-$800

OMC metal. $350-$400

Lighted Evinrude. $350-$500

Johnson. $350-$500

Great artwork.
$400-$600

Kawasaki.
Very popular.
$350-$500

Very rare Northway sign. $500-$800

Mercury Snowmobiles have a very strong collector base and race history. $600-$900

 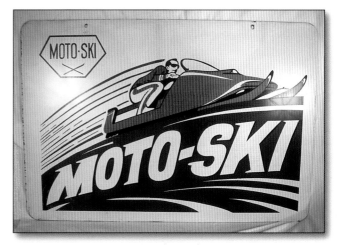

Lighted Moto-Ski. One of the best looking snowmobile signs ever made. $700-$1,000

1960s Moto-Ski lighted sign. $500-$800

Plain and simple Moto-Ski sign. $300-$500

Very old early 1960s Polaris sign. $600-$800

Impressive sign with strong colors and graphics. $500-$800

Polaris lighted sign. $400-$800

Small lighted Polaris sign. $250-$350

Rupp lighted sign. $400-$800

Very rare Polaron metal sign. $350-$500

Very early metal Scorpion sign. $500-$800

Late-1970s Scorpion sign. $300-$450

Small metal Scorpion sign. $250-$300

Early 1970s Scorpion sign. $400-$600

Rare outdoor wood AMF sign. $200-$400

Lighted AMF sign. $400-$600

Very obscure brand, lighted sign. $350-$500

1960s Ski-Doo metal sign. $750-$1,000

Ski-Doo lighted sign. $400-$800

Early 1970s metal sign. $350-$600

Lighted Skiroule with excellent graphics.
$400-$800

More simply designed Skiroule signs.
$300-$500

Metal Sno-Jet sign. $500-$900

Sno-Jet lighted sign. $400-$800

Very short-lived manufacturer signs. Sno-Hawk, Sno-Pony and Sno-Prince. $300-$500

Very rare Wildcat sign, nice artwork. $400-$600

Very popular Ski Whiz snowmobile signs. $600-$800

Yamaha lighted sign. $600-$1,000

Metal Yamaha sign. $400-$600

Early 1970's Yamaha small lighted dealer sign. $100-$250

Dealer Specific Items

Description: Many of the items in this section were only available to dealers for their showroom. These items are highly sought after by collectors because they are so limited and are easily the focal points of a collection.

Arctic Cat key FOB holder. $50-$100

Bosch dealer spark plug display. $40-$50

Arctic Cat dealer clock. Circa 1974. $100-$300

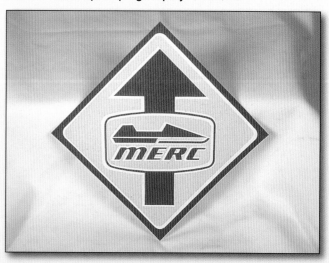
Mercury trail marker. $20-$30

Moto-Ski showroom sign. $50-$100

Moto-Ski poster. $100-$200

Hanging mobile. Moto-Ski. Very rare. $250-$500

Lighted roof light for vehicle. Moto-Ski. Unknown value.

Moto-Ski poster. $100-$200

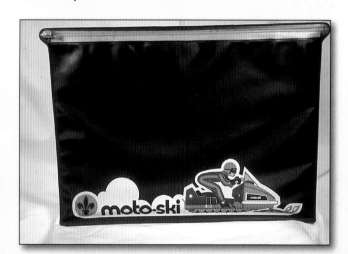

Dealer portfolios, Moto-Ski. $50-$75

Stuffed toy. Moto-Ski promotional item. $35-$50

Polaris thermometer. $100-$175

Dealer clock. Polaris. $100-$300

Dealer clock. Rupp. $100-$300

Personal phone from President of Scorpion.
Value unknown.

Ski-Doo placemat. $20-$50

1970s Ski-Doo trail marker. $20-$30

Original 1970s Ski-Doo dealer display of snowmobile goggles. $100-$175

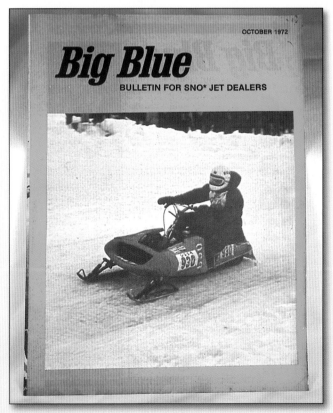

$20-$30

Thermometer, Union Carbide snowmobile oil. $50-$75

Dealer meeting promotional plate. Yamaha, 1981. $50-$150

Early 1970's Dealer banner. Gold and Blue. $100-$200

Yamaha dealer clock. $40-$75

Yamaha cardboard snowmobile. $50-$75

Early 1970's Dealer banner. Black/Red/White/Gold. $200-$350

1980 (left) and 1981 (right) Yamaha dealer wall poster displays. $100-$200

1971-72 Arctic Cat Dealer internal shop sign. Eight feet long. $150-$250

1971 Arctic Cat Dealer banner. Eight feet long. Very Rare. $400-$700

1970 Arctic Cat Dealer banner. Eight feet long. Very Rare. $500-800

Snowmobile Manufacturer's Advertising Posters

Description: Vintage posters and banners such as these pictured in this section are highly sought after by collectors because of the large artwork. Posters are always a popular "snowmobile garage" addition. Most posters are valued between $50-$100. Premiums are paid for pristine condition examples.

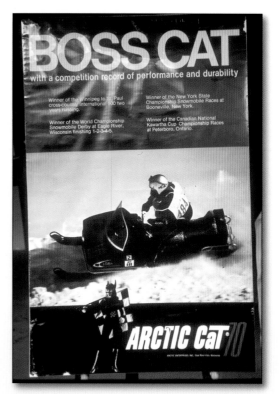

Early 1960s Arctic Cat poster. $150-$250

1970 Arctic Cat poster. Very popular. $75-$150

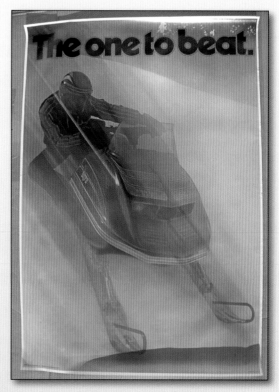

1974 Arctic Cat El Tigrè posters. $50-$85

1971 Arctic Cat poster. $85-$125

1971 Arctic Cat poster. $85-$125

Kawasaki poster, late 1970s. $40-$70

Hard to find Columbia poster from early 1970s. $40-$60

Moto-Ski poster. $20-$30

Rare Polaris Snow Traveler poster. Late 1950s. $150-$200

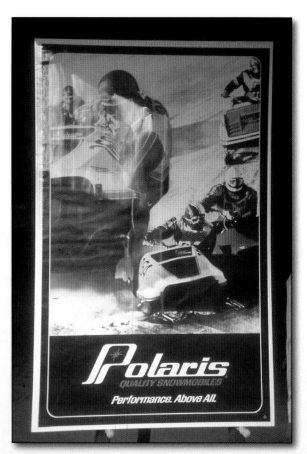

1970s Polaris Posters. $50-$100

1970s Polaris Posters. $50-$100

Polaris banners, 1970s. $150-$300

1970s Polaris posters. $85-$125

Raider banner, early 1970s. Very rare. $100-$300

Early 1970s Rupp banner. $200-$400

Scorpion banner. Very Rare. $200-$400

Mid-1970s Scorpion banner. $200-$300

Very popular Scorpion racing poster.
$85-$125

Pictures the popular RV. $100-$175

Not very graphic. $50-$75

Small size. $25-$50

Ski-Doo dealer poster circa 1974. $85-$125

Ski-Doo poster. 1970s. $85-$125

Early 1970s Speedway banner. $250-$400

Starcraft banner. Very rare. $100-$200

Yamaha dealer show room window posters. Clear material. Very rare. 4 feet by 8 feet. $150-$250

Yamaha SRX poster 1981. Very collectible. $85-$125

Yamaha SRX 500 promotional picture 8"x 11" Very rare. Value unknown.

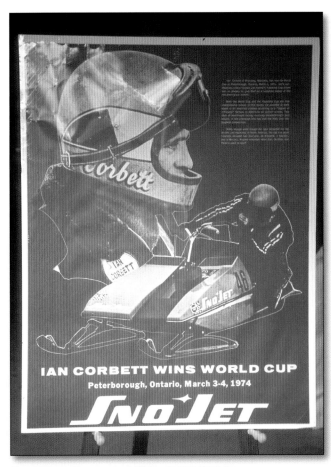

Sno-Jet poster early 1970s. $100-$200

Quaker State Oil poster. Excellent artwork. $75-$100

Skiroule poster, collector favorite because of bathing suit model on the snowmobile, circa 1975. $100-$200

Chapter 6
Vintage Advertising

One aspect of vintage snowmobile collecting that brings smiles to collector's faces is looking through old snowmobile magazines. Not only do the articles bring a sense of nostalgia to the reader, but the advertisements many times are the most interesting parts of the magazines.

Many collectors actually will cut out the advertisements and put them in three ring binders for easy viewing. What makes vintage advertisements important to collecting vintage snowmobiles and collectibles? The importance lies in what is pictured in the old advertisements. The collector can finally have the portal to the past by looking at how the sleds looked when new, what attire the manufacturer was promoting for the specific year, in addition to some accessories which may be depicted.

Not to be forgotten, is the fact that many times a modern snowmobiler can hardly believe people rode these olds sleds in the manner depicted. Many vintage advertisements show riders with no helmet, no hat, stylish snow suits, all pictured in the dead of winter while riding fast on the new snowmobile and sold as family fun!

The collector can find these old advertisements fairly easy inside old snowmobile magazines. Many of these magazines can be found for under $10 each. However, sometimes these magazines can fetch $30 or more depending on what is the main feature in the magazine. This chapter illustrates some of the covers of old snowmobile magazines and many of the different advertisements from some of the brands that were in existence through the Golden Years of snowmobiling.

In addition to advertising that was done on paper, this chapter is also devoted to all the other snowmobile advertising items that are not easily categorized such as all sorts of odd items from records, cups, playing cards, ash trays, cufflinks, stir sticks and many items you probably didn't even know were made.

The manufacturer's and other companies who produced snowmobile related items, many times made items that were promotional items, giveaways, or licensed merchandise that bore the manufacturers name. Most of these small collectibles are very hard to find and were very easy to loose, or become obsolete and thrown out. However, some items have stood the test of time and can still be found at garage and estate sales just because they were so small and could be squirreled away in basement trunks and junk drawers.

Literature and Magazine Advertisements

Many vintage magazine and literature pieces are highly sought out by collectors. This is just a small sampling of some paper items. A separate book could be compiled on all the collectible literature and magazines that have been produced in the past 30-plus years. Prices can range from $5 - $50 depending on the rarity and particular issues contents.

$5-$10

$15-$30

$15-$30

$15-$30

$20-$50 Very hard to find.

$5-$20

Various magazines from the 1970s-1982.

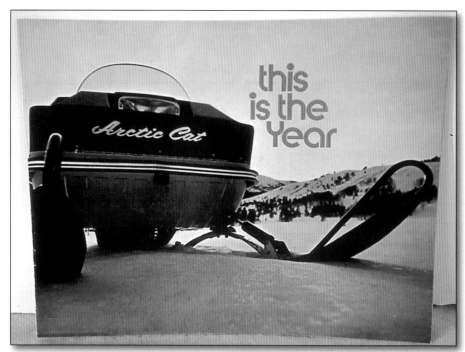

1971 Arctic Cat brochure.
$25-$40

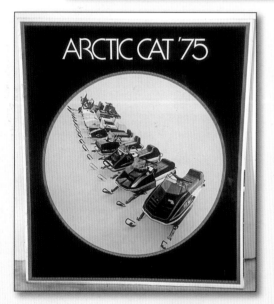

1970 Arctic Cat brochure.
$25-$40

Many manufacturer's literature pieces can command very high prices.

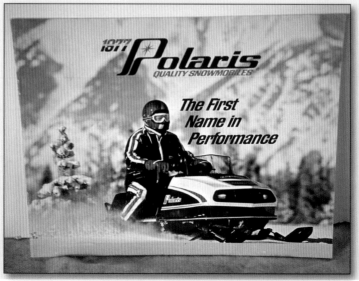

1977 Polaris brochure.
$15-$25

1978 Yamaha Snowmobiles

Yamaha brochures from 1978, 1980 and 1981. $15-$30

1978 Polaris RXL Sno-Pro operator's manual. Rare. $50-100

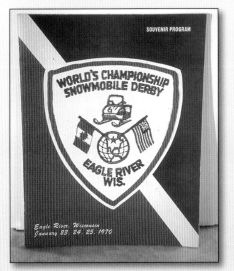

Early 1970s Eagle River Snowmobile Derby programs, very rare. $20-$50

Smoking Related Advertising Items

During the past fifty years smoking has evolved from a habit that the majority of people did to a small percentage participating nowadays. The manufacturer's were quick to brand smoking paraphernalia and some manufacturer's even supported the habit by installing cigarette lighters on some of their luxury snowmobiles such as the 1975 Arctic Cat Pantera.

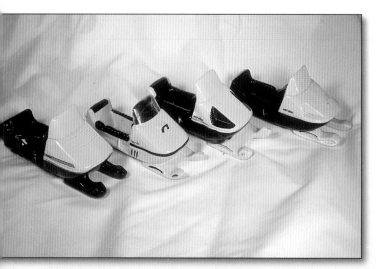

Very hard to find Auto Ski ash tray. $20-$40

Very collectible Ski-Doo ash tray with snowmobile affixed. $30-$50

Snowmobile ash trays. Brand specific ask trays can sell for $10-$40 depending on brand and condition.

Moto-Ski ash tray. $10-$25

Very collectible Scorpion ash trays. $20-$40

Ceramic ash trays. $15-$25

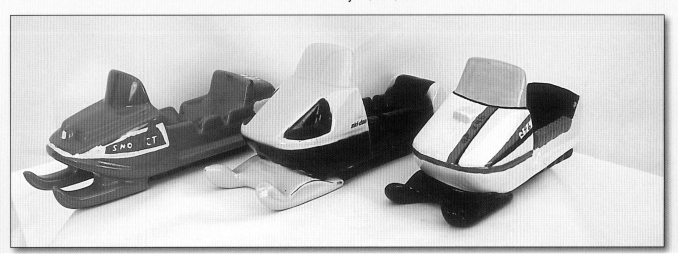

Zippo-style lighters. Highly collected by snowmobile enthusiasts and lighter collectors. $40-$75

Highly collectible Zippo lighter with Arctic Cat artwork. $75-$125

Hard to find Yamaha lighter. $50-$85

Advertising Playing Cards

Before the onslaught of 200 plus cable channels, the Internet, and electronic games, people actually played games with friends and family. Many manufacturer's created and branded playing cards and games for this purpose. Values range between $25-$50.

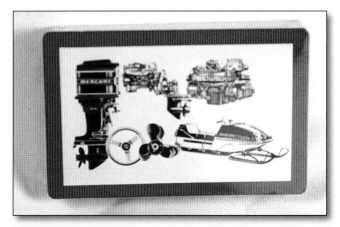

Mercury playing cards.

Ski Whiz playing cards with rare can opener premium. $35-$50

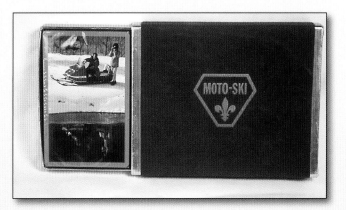

Moto-Ski.

Sno-Jet. $35-$50

Ski-Doo.

John Deere.

Drinking Containers and Related Advertising Items

During the golden age of snowmobiling most snowmobiles did not go very fast and one of the social activities was to gather up family, friends or snowmobile club members and go for a snowmobile ride. This usually meant going from one bar to another and in many regions of the snow belt states this could mean many bars in a very short distance. Drinking and driving was not a big concern during those days because the snowmobiles did not go very fast and the distances traveled were short.

Many of the snowmobile manufacturers were quick to realize that snowmobilers were very brand loyal and that discussions of reliability and performance were often the topic of discussion at these bars. So, what better place to put your brands name but on the glass, stir sticks and coasters in the bars, or for sale for use at home.

Collectible drinking glasses. $20-$40

Miscellaneous Ski-Doo promotional items. $15-$25

1971 Arctic Cat mug. $30-$40

1977 Yamaha promotional pewter mug with see-through bottom. $30-$50

Very rare Polaris four glass set. $80-$100

Scorpion glasses and Styrofoam cup. Very collectible. $20-$40

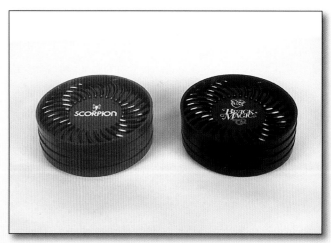

Scorpion drink coasters. $10-$15

Portable alcoholic beverage container. $25-$50

Ezra Brookes whiskey decanter. $25-$50

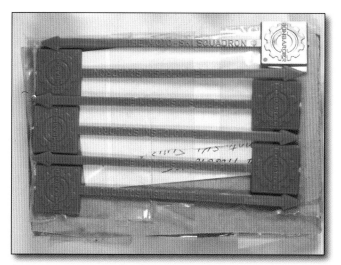

Ski-Doo stir sticks. $3-$5

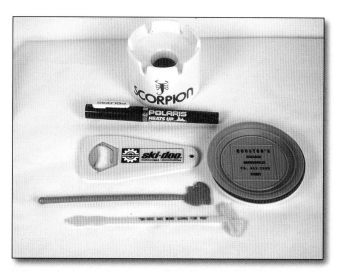

Different companies promotional items. $15-$30

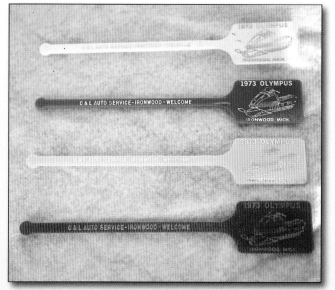

Early 1970s Ironwood Olympus drink stir sticks with Ski-Doo pictured. Very rare. $15-$30

Scorpion drink stir stick. $15-$30

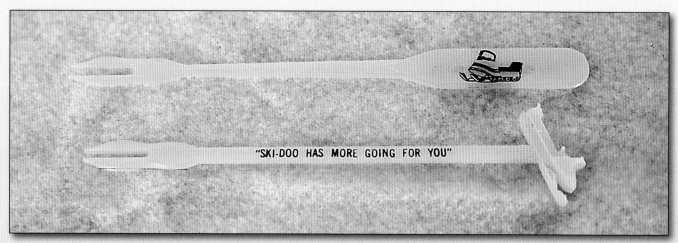

Ski-Doo drink stir sticks. Very rare. $15-$30

Large beer advertising poster. $50-$75

Schlitz beer sign. Very rare: $150-$300

Scorpion branded, 86 proof Blended Whiskey, circa late-1960s. Extremely rare unopened. $150-$250

Miscellaneous Advertising Items

Many snowmobile collectibles are unique and do not easily fall into one particular collectible category. Included in this section are some of those items.

Skiroule key chain and cuff links. $25-$40

1970s John Deere key chain, Sno-Jet cuff links and clip, Skiroule clip, Yamaha earrings. $25-$40

Vintage 1970s belt buckles by several brands. $25-$40

Key chains by Sno-Jet, Northway, and John Deere. $25-$40

Arctic Cat cuff links and tie tack. Circa early 1970s. $25-$40

Yamaha salt and pepper shakers, mid-1970s. $25-$50

Collector buttons. $5-$10

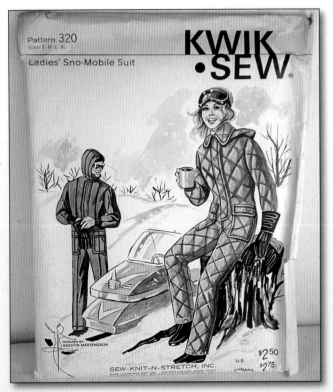

Make your own snowmobile suit (don't forget to add patches).

Arctic Cat repair patch kit, 1970s. Nice artwork. $25-$35

Yamaha collector plate. $50-$100

Yamaha post card, early 1970s. $5-$10

1970s Arctic Cat burglar alarm, rare. $80-$150

Arctic Cat lawn mower. Some collectors collect everything made by a particular manufacturer. This is a very hard to find item. $75-$150

Cat Klaws. Popular traction product for cleated track snowmobiles. $15-$25

Spark plug caddies. $10-$20

Very rare early Ski-Doo snow shoe accessory, branded Ski-Doo. $150-$250

1970s advertisement to win a Ski-Doo. You can just imagine this sitting on the bar counter in 1970. $50-$85

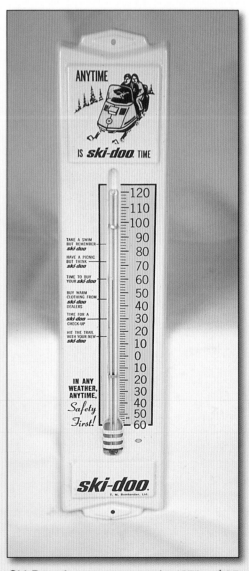

Ski-Doo thermometer, early 1970s. $85-$125

7-Up soda can with snowmobile on can. $10-$15

Plaster-cast snowmobile with boy rider. Sold in craft stores for customized painting. Depending on quality of paint work, $50-$100.

Schmidt Beer can with 1972 Arctic Cat Panther displayed. $10-$20

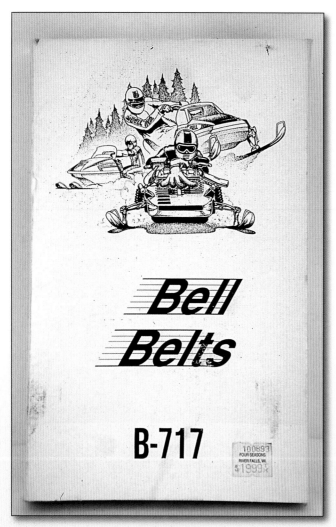

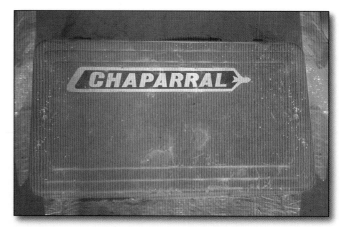

Vintage 1970s Chaparral tool kit. $40-$85

Vintage snowmobile replacement belt sleeve (collectors buy stuff for the cool artwork and this is representative).

Chaparral saddle bags. Very rare. $100-$150

Moto-Ski kidney belt. Ski-Doo also made one too. For the long ride on the old bogie wheel suspension! $35-$50

$20-$40

$20-$40

Different snowmobile aftermarket parts items. Value is driven by the artwork rather than the item.

$20-$40

$35-$50

Avon soap. Very hard to find. $25-$40

Avon cologne in snowmobile shape. $25-$40

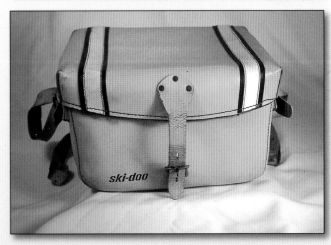

Early 1970s Ski-Doo saddlebags. $100-$150

John Deere stained glass lamp with snowmobile depicted. Very rare. $150-$300

1972 Arctic Cat gym bag. $40-$85

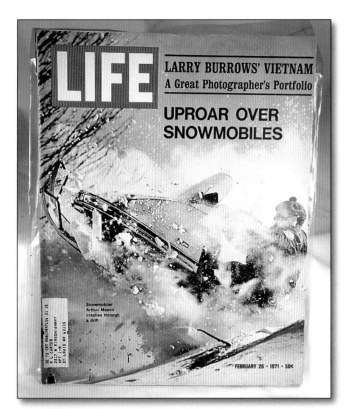

LIFE magazine with featured cover and inside article on snowmobiling. $10-$20

Assortment of snowmobile coloring books. $15-$25

Snowmobile comic books for the kids. $10-$25

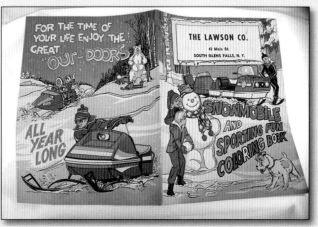

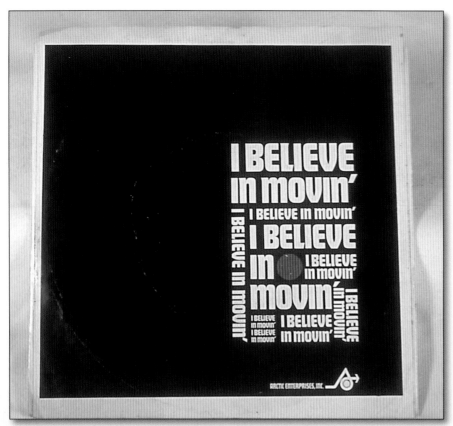

Arctic Cat 45 rpm record. $25-$40

Not only was snowmobiling fun, but singing snowmobile songs at the bar or at home was even promoted.

1971 Arctic Cat promotional 45 rpm record.

Good cover art drives value up. $35-$50

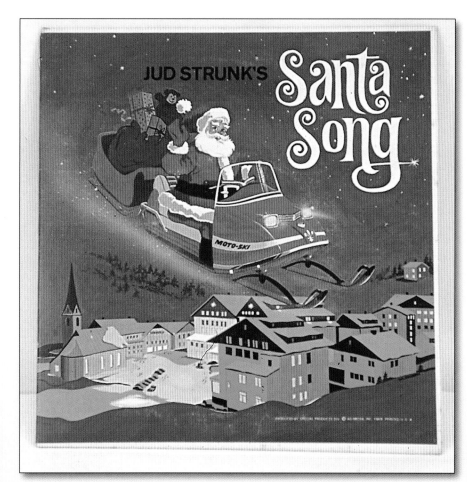

Sno-Jet record. $25-$40

Snowmobiling was such a hit in the late 1960s and 1970s that records and songs were made either singing or talking about tales of snowmobiling, or the manufacturers brand, or just of the engine noise of snowmobiles!

John Deere record. $25-$40

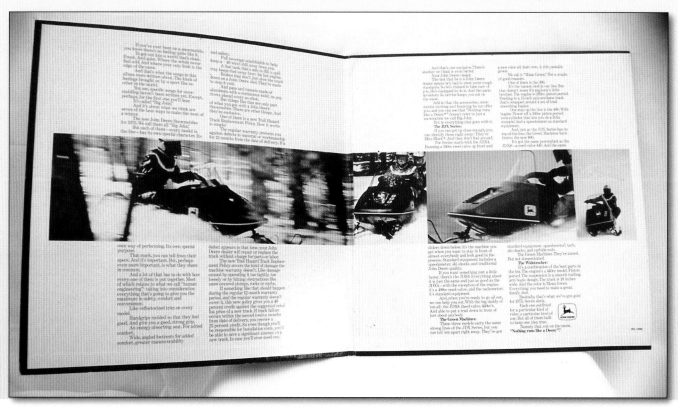

8 OZ

1969 Kellogg's Corn Flakes Cereal Box, unopened with special promotioin to win a first prize of two OMC Snowcruisers in the Deluxe trim package with a 440 engine and electric start. Second prize was 23 Model 200 Snowcruisers. Extremely rare and offered only in Canada. $100-$200

More great books from Iconografix

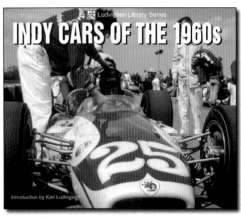
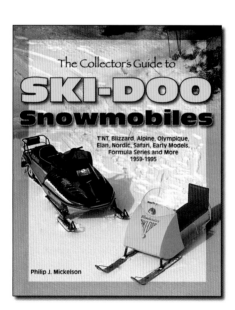

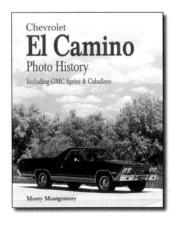
Iconografix, Inc.
P.O. Box 446, Dept BK,
Hudson, WI 54016
For a free catalog call: 1-800-289-3504
info@iconografixinc.com
www.iconografixinc.com